Viola Kress

Aus der Reihe: e-fellows.net stipendiaten-wissen

e-fellows.net (Hrsg.)

Band 1405

Should Genetically Modified Foods Be Included As a Part of The Human Diet and Food Industries in Australia?

A Review of Possible Benefits and Health Risks

GRIN Publishing

Bibliographic information published by the German National Library:

The German National Library lists this publication in the National Bibliography; detailed bibliographic data are available on the Internet at http://dnb.dnb.de .

Imprint:

Copyright © 2010 GRIN Verlag GmbH
Print and binding: Books on Demand GmbH, Norderstedt Germany
ISBN: 978-3-656-97359-1

This book at GRIN:

http://www.grin.com/en/e-book/299987/should-genetically-modified-foods-be-included-as-a-part-of-the-human-diet

GRIN - Your knowledge has value

Since its foundation in 1998, GRIN has specialized in publishing academic texts by students, college teachers and other academics as e-book and printed book. The website www.grin.com is an ideal platform for presenting term papers, final papers, scientific essays, dissertations and specialist books.

Visit us on the internet:

http://www.grin.com/

http://www.facebook.com/grincom

http://www.twitter.com/grin_com

Should genetically modified foods be included as a part of the human diet and food industries in Australia?

—

A review of possible benefits and health risks

Author: Viola Kress

Date of submission: 09.09.2010

Table of contents

1. Introduction

The development of genetically modified (GM) or transgenic foods promises another agricultural revolution, producing more and high-quality food to feed the growing world population (Uzogara 2000). Within a rather short period of time genetic modifying has evolved from basic science to commercial applications (Engel, Frenzel & Miller 2002). According to recent statistics (ISAAA 2009), transgenic crops are cultivated in 25 countries on a hectarage of approximately 134 million hectares, an 80-fold increase compared to 1999. However, their introduction into the human food supply activated an apparently never ending debate about safety, potential risk and ethical concerns (FSANZ 2005; Better Health Channel 1999). According to Trojanowicz, Latoszek and Poboży (2010) the sources of these controversies can be related to consumer suspicions regarding the true intentions of biotechnological producers and mistrust in government bureaucracies.

The aim of this literature review is to analyse current scientific research about GM foods with regard to the question whether GM food should be introduced in Australia or not. Part 1 includes a definition of GM foods and a description of current commercialized GM crops worldwide. The second part provides an overview of possible advantages and disadvantages concerning GM foods. Therefore, it identifies the main groups which profits from GM crops. This review ends with a conclusion and a definition of the author's position concerning the controversial issue of including GM foods as part of the Australian diet and food industries.

2. Definition and current availability of GM foods

According to the Food Standards Australia New Zealand (2005), a GM food is a food which has been derived or developed from a GM organism whose genome has been modified by genetic engineering techniques. Gene technology uses special techniques like copying and transfer of genes from one organism to another to alter the genetic material (DNA) of plants, animals or microorganisms (FSANZ 2005). Hence, genetic engineering may generate organisms with a much broader genetic diversity and additional or modified traits through combining genes of different species (Houdebine 2010; Trojanowicz, Latoszek & Poboży 2010; Batista & Oliveira 2009). The steps involved in developing a GM organism are shown in Appendix A. Furthermore the expression 'GM foods' includes foods containing GM ingredients as well as food additives and processing aids which were produced by using genetic modification (FSANZ 2005).

The main traits of all GM crops available on the international market are resistance to insects and tolerance to certain herbicides (FSANZ 2005). Table 1 shows the current producers of GM crops, the types of commercialized GM crops and their global hectarage.

Table 1. Global Area of Biotech Crops in 2009: by Country (Million Hectares)

Rank	Country	Area (million hectares)	Biotech Crops
1*	USA*	64.0	Soybean, maize, cotton, canola, squash, papaya, alfalfa, sugarbeet
2*	Brazil*	21.4	Soybean, maize, cotton
3*	Argentina*	21.3	Soybean, maize, cotton
4*	India*	8.4	Cotton
5*	Canada*	8.2	Canola, maize, soybean, sugarbeet
6*	China*	3.7	Cotton, tomato, poplar, papaya, sweet pepper
7*	Paraguay*	2.2	Soybean
8*	South Africa*	2.1	Maize, soybean, cotton
9*	Uruguay*	0.8	Soybean, maize
10*	Bolivia*	0.8	Soybean
11*	Philippines*	0.5	Maize
12*	Australia*	0.2	Cotton, canola
13*	Burkina Faso*	0.1	Cotton
14*	Spain*	0.1	Maize
15*	Mexico*	0.1	Cotton, soybean
16	Chile	<0.1	Maize, soybean, canola
17	Colombia	<0.1	Cotton
18	Honduras	<0.1	Maize
19	Czech Republic	<0.1	Maize
20	Portugal	<0.1	Maize
21	Romania	<0.1	Maize
22	Poland	<0.1	Maize
23	Costa Rica	<0.1	Cotton, soybean
24	Egypt	<0.1	Maize
25	Slovakia	<0.1	Maize

* 15 biotech mega-countries growing 50,000 hectares, or more, of biotech crops

Source: Clive James, 2009.

Source: Clive 2009 in ISAAA 2009

At present, certain varieties of cotton and two versions of canola are the only permitted GM crops grown in Australia (FSANZ 2005). The FSANZ has also approved certain varieties of other GM foods like soya bean, corn, sugar beet and potato which can be imported and sold in Australia for human consumption (FSANZ 2005). These foods are commonly present in breads, pastries, snack foods, baked products, oils and fried foods (Carman 2004; Better Health Channel 1999).

Significantly, the current labelling laws exclude foods which are produced from animals fed with GM feed such as milk or meat, which are prepared at the point of sale like in restaurants as well as highly refined foods which contain no DNA or Protein like cooking oils or sugars (IHER 2008; FSANZ 2005). Furthermore these regulations do not cover foods contaminated accidentally by up to one per cent per ingredient nor those which have been processed before 7 December 2002 or which contain GM flavours present at less than one per cent (IHER 2008; FSANZ 2005). The production of foods with processing aids or food additives using GM microbes is also ignored (IHER 2008; FSANZ 2005).

3. Benefits of introducing GM foods into Australia

It appears that GM food can be classified into two groups (Celec et al. 2005 in Trojanowicz, Latoszek & Poboży 2010), or alternatively into three (Magaña-Gómez & Calderón de la Barca 2009).

The first generation has been biotechnologically derived to create various improvements in the production such as resistance to pests and herbicides (Magaña-Gómez & Calderón de la Barca 2009). Therefore, GM foods are potentially better for the environment. The amount of pesticides, for example, could be reduced by developing crops which are tolerant to particular herbicides like glyphosate (Better Health Channel 1999). This results in lower soil erosion, moisture loss and fuel costs because of less tillage practices and fewer herbicide applications (Brookes & Barfoot 2006 in Batista & Oliveira 2009). However, a case study of Zhou & Kastenberg (2006) in the United States identified an increase of human health risks since the introduction of herbicide-tolerant (HT) crops associated with huge glyphosate applications. In addition, Benbrook (2004) and Ho (2010) state that the use of herbicides is encouraged by HT crops, leading to herbicide-resistant weeds which require yet more herbicides.

Not only HT, but also pest and disease-resistant crops count among the first generation and may contribute to the reduction of pesticides (Goodyear-Smith 2001; TechNyou 2010). Consequently, the chemical pollution of the environment can be reduced and less pesticide residue will be present in air, water, soil and foods (Goodyear-Smith 2001; Cerdeira & Duke 2006). This may also initiate an improvement of the health and safety for farmers and farm workers like it occurred in China according to a three-year-survey from 1999 (Pray et al. 2002). Conventional cotton, for example, is very susceptible to insect harm and 15% of the US insecticide production is used on this one crop (Halford & Shewry 2000). In contrast, Bt[1] cotton requires only 15% of the insecticide used on conventional cotton (Halford & Shewry 2000). Moreover, the introduction of insect-resistant crops may lead to a greater continuity of food supply and lower prices (Goodyear-Smith 2001). In the UK, for example, slow-ripening

---

[1] Bt cotton = cotton which contains the gene for insect toxin production of the bacterium Bacillus thuringiensis (Goodyear-Smith 2001).

GM tomatoes are cheaper than their non-GM competitors because of reduced processing costs like (Halford & Shewry 2000).

The second, also called new generation consists of crops with new traits orientated toward consumers. It claims to offer benefits such as increased levels of proteins and carbohydrates, modified and healthier fats, improved flavour characteristics, or increased levels of micronutrients or other phytochemicals (Magaña-Gómez & Calderón de la Barca 2009). For example, GM golden rice containing the vitamin A gene from a daffodil plant has been developed (Ye et al. 2008; Better Health Channel 1999). Therefore, further modification could be used in the future to create GM food crops with increased nutritional value and higher amounts of nutrients like iron which is the most common nutritional deficiency in Australia (FSANZ 2005; FSANZ 2006; Ye et al. 2008; Batista & Oliveira 2009).

A third generation of GM plants is in development which should have a greater tolerance to resist abiotic stress such as drought, high temperatures, or saline-polluted soils (Goodyear-Smith 2001; FSANZ 2005; Magaña-Gómez & Calderón de la Barca 2009). It also includes plants which are used as biological production systems for manufacturing pharmaceutical compounds such as anticancer agents (Magaña-Gómez & Calderón de la Barca 2009; Batista & Oliveira 2009).

4. Potential problems associated with eating GM foods

Even though there may be a lot of advantages of adopting GM foods, public concerns have arisen about the possible unexpected new risks. According to Améndola et al. (2006), the hype about the benefits of GM crops is mainly the result of propaganda by the biotech industry and industry-sponsored organizations, whereas negative impacts and problems are often ignored.

The introduction of GM foods was in 1994, therefore no studies of long-term effects on human health have been carried out (TechNyou 2010; Batista & Oliveira 2009). One risk associated with GM foods is the potential appearance of toxins, allergens or genetic hazards as a consequence of the biosynthesis and accumulation of new specific chemical metabolites in the human diet (Conner & Jacobs 1999; Goodyear-Smith 2001). GM is unpredictable due to possible interaction of genes (Dona & Arvanitoyannis 2009). For instance, the transfer of allergens from traditional foods into GM foods could be possible. In 1996, a GM soya bean was developed including a gene from Brazil nut which codes for a methionine-rich protein (Goodyear-Smith 2001; Batista & Oliveira 2009). However, the development of this crop was stopped after finding allergens in the GM soya bean. The majority of commercialized GM foods are developed by introducing genes from sources with unknown allergenic potential (Batista & Oliveira 2009). Thus the resulting products have a potential risk of allergenicity.

Even though, no cases of food toxicity resulting from GM food have been occurred yet, several incidents have raised public concerns like the case of the company Showa Denko from 1989 (Carman 2004). The production of the amino acid tryptophan from a GM strain of *Bacillus amyloliquefacien* resulted in an epidemic of eosinophilia myalgia syndrome in the United States and Europe with 37 deaths and more than 1,500 non-fatal reported cases (Goodyear-Smith 2001; Carman 2004). However, the company destroyed their stocks of the GM bacterium strain before the examination could be completed. Therefore, it was not possible to establish whether the use of the GM bacteria or an inadequate purification process cased the epidemic. Furthermore, a study by Ewen and Pustzai (1999 in Batista & Oliveira 2009) reported adverse effects in the gut of rats after feeding them with GM potatoes, yet this experiment was criticized afterwards due to methodically deficiencies.

Another potential unintended effect is the transfer of new genetic material from the GM food to bacterial cells in the human gastrointestinal tract (FSANZ 2005). Human in vitro simulations by Martín-Orúe et al. (2002) and research by Netherwood et al. (2004) indicated that transgenic DNA in GM food may survive in the human stomach and the small intestine for up to four hours. Consequently, not all DNA present in food is digested in the small bowel which may lead to gene flow. Furthermore, a study from Gebhard and Smalla (1998) showed that bacteria can absorb and integrate DNA fragments derived from GM sugar beet, albeit under idealized and artificial laboratory conditions. In addition, several animal studies evidenced that plant and gastrointestinal DNA can be transferred to mammalian and bacterial cells (Einspanier et al. 2001; Schubbert, Lettmann & Doerfler 1994). On the other hand, Flachowsky (2005 in Dona & Arvanitoyannis 2009) state that the absorption of GM DNA into cells of the gastrointestinal tract will usually not exert any biological effect.

Moreover, a few reports have raised concerns regarding the potential transfer of antibiotic resistance (AR) genes. Bioengineers sometimes use these genes in GM plants as a marker for identification of successful gene transfer (Better Health Channel 1999; Goodyear-Smith 2001). If such AR genes from GM foods entered the food chain and were transferred to pathogenic microorganisms in the human digestive system, this may reduce the effectiveness of relevant antibiotics in the treatment of human disease and increase human infection risk (Better Health Channel 1999; Goodyear-Smith 2001). Van Den Eede et al. (2004) maintain that AR gene transfer may occur rarely due to different barriers within the gastrointestinal tract such as the existence of bacterial restriction enzymes. Developing GM

5

plants without the AR gene is sometimes possible, but expensive and time-consuming (Batista & Oliveira 2009).

In addition, there are some concerns about ethical, religious and social issues concerning the introduction of GM food. Eating foods which are modified by transferring genetic material from animals to plant foods may be a problem for certain cultural and religious groups like Muslims (Uzogara 2000; Better Health Channel 1999). Vegetarians may similarly object to vegetables and fruits that contain any animal genes (Crist 1996 in Uzogara 2000). However, there is no GM crop commercialized containing animal genes at present.

5. Main profiteers of GM crops

Biotechnology and agribusiness companies are among the biggest profiteers of GM crops. The introduction of GM crops has been dominated and promoted by a few companies, including Monsanto, Syngenta, and Bayer which leads to unequal trade advantages because of patent restriction with high technology fees (Améndola et al. 2010; Klingeman & Hall 2006).

Furthermore, it is often claimed that GM crops may bring financial and environmental benefits to farmers. The development of crops which are resistant to pests, herbicide, weather and other environmental stresses could lead to a reduction of crop loss and may increase the crop yield (Uzogara 2000). Furthermore, the decreased reliance on pesticides may reduce costs, environmental pollution and adverse health effects as a result of less exposure to high amounts of dangerous pesticides. In fact, the advantages of GM crops for farmer are unclear and farmers achieve only moderate benefits at present (Klingeman & Hall 2006; Wu 2004). In addition, Améndola et al. 2010 point out that small farmers in several developing countries are expelled from their lands by large landowners to expand the cultivation of crops, mainly GM.

It is also suggested that lower processing costs may lead to a price decrease for GM food compared to conventional food which could be a benefit for the consumers. Although it is expected that consumers could gain particularly from the second and third generation of GM crops because of their possible higher nutritional contents and quality (Batista & Oliveira 2009), Wu (2004), Halford and Shewry (2000) state that the welfare gain to individual consumers is small and may not outweigh perceived risks. Moreover, improved nutritional qualities and agronomic properties may provide considerable benefits in developing countries, but not in western societies according to Goodyear-Smith (2001) and Engel, Frenzel and Miller (2002).

6. Conclusion

Even though this report has presented some possible advantages of GM foods and crops and demonstrated that the probability of GM foods causing adverse health effects is low, it cannot be guaranteed that genetically engineered foods are safe at present. The consequences of exposing millions of people to it, including the whole of the Australian population, could be risky. Hence, further scientific effort and research is necessary, especially about long-term validation of safety issues, in order to build confidence in the evaluation and acceptance of GM foods. While there are currently no standardized and adequate methods to evaluate the safety of GM foods, attempts towards equalization are in process. However, until these methods have been adopted, the GM foods should be excluded from the Australian diet and food industries. Under current labelling regulations it is not required to indicate if GM foods are present and therefore consumers do not have a real choice. Thus, it would be necessary to label GM foods sufficiently and completely to enhance the consumer's position.

List of References

Améndola, C, Pereira, M, Sánchez, J, Mayet, M, Bebb, A, Freese, B & López, J 2006, *Who benefits from GM crops?*, viewed 2nd September 2010, Friends of the Earth, <http://www.foe.co.uk/resource/reports/who_benefits_from_gm_crops.pdf>.

Batista, R & Oliveira, MM 2009, 'Facts and fiction of genetically engineered food', *Trends in Biotechnology*, vol.27, no.5, pp 277-286, viewed 4th September 2010, Scopus.

Benbrook, CM 2004, *Genetically engineered crops and pesticide use in the United States: The first nine years*, viewed 2nd September 2010, BioTech InfoNet, < http://www.biotech-info.net/Full_version_first_nine.pdf>, GoogleScholar.

Better Health Channel 1999, *Genetically modified food*, viewed 3rd September 2010, State Government of Victoria, <http://www.betterhealth.vic.gov.au/bhcv2/bhcarticles.nsf/pages/Genetically_modified_food>.

Carman, J 2004, 'Is GM food safe to eat?', in R Hindmarsh & G Lawrence (eds.), *Recoding Nature: Critical Perspectives on Genetic Engineering*, University of New South Wales Press, Sydney, pp 82-93.

Cerdeira, AL & Duke, SO 2006, 'The current status and environmental impacts of glyphosate-resistant crops: A review', *Journal of Environmental Quality*, vol.35, no. 5, pp 1633-1658, viewed 7th September 2010, ProQuest.

Conner, AJ & Jacobs, JME 1999, 'Genetic engineering of crops as potential source of genetic hazard in the human diet', *Mutation Research/ Genetic Toxicology and Environmental Mutagenesis*, vol.433, no.1-2, pp 223-234, viewed 22nd August 2010, ScienceDirect.

Dona, A & Arvanitoyannis, IS 2009, 'Health risks of genetically modified foods', *Critical Reviews Food Science and Nutrition*, vol.49, no.2, pp 164-175, viewed 9th September 2010, Informaworld.

Einspanier, R, Klotz, A, Kraft, J, Aulrich, K, Poser, Schwaegele, F, Jahreis, G & Flachowsky, G 2001 'The fate of forage plant DNA in farm animals: a collaborative case-study investigating cattle and chicken fed recombinant plant material', *European Food Research Technology*, vol. 212, no.2, pp129-134, viewed 8th September 2010, SpringerLink.

Engel, K-H, Frenzel, T & Miller, A 2002, 'Current and future benefits from the use of GM technology in food production', *Toxicology Letters*, vol.127, no.1-3, pp 329-336, viewed 4th September 2010, Scopus.

Food Standards Australia New Zealand (FSANZ) 2005, *Safety assessment of genetically modified foods*, viewed 27th August 2010, <http://www.foodstandards.gov.au/_srcfiles/GM%20Foods_text_pp_final.pdf>.

Food Standards Australia New Zealand (FSANZ) 2006, *Ferric sodium edetate as a permitted form of iron*, viewed 7th September 2010, <http://www.foodstandards.gov.au/_srcfiles/IAR_A570_Ferric_Sodium_Edetate.pdf>.

Gebhard, F & Smalla, K 1998, 'Transformation of Acinetobacter sp. Strain BD413 by Transgenic Sugar Beet DNA', *Journal of Applied and Environmental Microbiology*, vol.64, no.4, pp 1550-1554, viewed 5th September 2010, American Society for Microbiology, <http://aem.asm.org/>.

Goodyear-Smith, F 2001, 'Health and safety issues pertaining to genetically modified foods', *Australian and New Zealand Journal of Public Health*, vol.25, no.4, pp 371-375, viewed 30th August 2010, Scopus.

Halford, NG & Shewry, PR 2000, 'Genetically modified crops: methodology, benefits, regulation and public concerns', *British Medical Bulletin*, vol.56, no.1, pp 62-73, viewed 5th September 2010, Oxford University Press Journals.

Ho, M-W 2010, *GM crops facing Meltdown in the USA*, viewed 7th September 2010, Institute of Science in Society, < http://www.i-sis.org.uk/GMCropsFacingMeltdown.php>.

Houdebine, L-M 2010, 'Animals and plants: Genetic modification', *Encyclopedia of Biotechnology in Agriculture and Food*, viewed 6th September 2010, Informaworld.

Institute of Health and Environmental Research (IHER) 2008, *Is GM food safe to eat?*, viewed 1st September 2010, <http://www.iher.org.au/publications.php?pubID=3>.

International Service for the Acquisition of Agri-Biotech Application (IAAA) 2009, *Global status of commercialized Biotech/ Gm crops 2009*, viewed 8th September 2010, <http://www.isaaa.org/resources/publications/briefs/41/default.asp>.

Klingeman, WE & Hall, CR 2006, 'Risk, trust, and consumer acceptance of plant biotechnology', *Journal of Crop Improvement*, vol.18, no.1, pp 451-486, viewed 8th September 2010, Scopus.

Magaña-Gómez, JA & Calderón de la Barca, AM 2009, 'Risk assessment of genetically modified crops for nutrition and health', *Nutrition Reviews*, vol.67, no.1, pp 1-16, Wiley InterScience.

Martín-Orúe, SM, O'Donnell, AG, Arino, J, Netherwood, T, Gilbert, HJ & Mathers, JC 2002, 'Degradation of transgenic DNA from genetically modified soya and maize in human intestinal stimulations', *British Journal of Nutrition*, vol.87, no.6, pp 533-542, viewed 4th September 2010, ProQuest.

Netherwood, T, Martín-Orúe, SM, O'Donnell, AG, Gockling, S, Graham, J, Mathers, JC & Gilbert, HJ 2004, 'Assessing the survival of transgenic plant DNA in the human gastrointestinal tract', *Nature Biotechnology*, vol.22, no.2, pp 204-209, viewed 28th August 2010, Scopus.

Pray, CE, Huang, J, Hu, R & Rozelle, S 2002, 'Five years of Bt cotton in China – the benefits continue', *The plant journal*, vol.31, no.4, pp 423-430, viewed 7th September 2010, Scopus.

Schubbert, R, Lettmann, C & Doerfler, W 1994, 'Ingested foreign (phage M13) DNA survives transiently in the gastrointestinal tract and enters the bloodstream of mice', *Molecular and General Genetics*, vol.242, no.5, pp 495-504, viewed 8th September 2010, SpringerLink.

TechNyou 2010, *Arguments for and against gene technology*, viewed 27th August 2010, < http://technyou.edu.au/wp-content/uploads/2009/08/ArgForAgainstMay102.pdf>.

Trojanowicz, M Latoszek, A & Poboży, E 2010, 'Analysis of genetically modified food using high-performance separation methods', *Analytical Letters*, vol.43, no.10, pp 1654-1679, viewed 6th September 2010, Informaworld.

Uzogara, SG 2000, 'The impact of genetic modification of human food in the 21st century: A review, *Biotechnology Advances*, vol.18, no.3, pp 179-206, viewed 5th September 2010, Scopus.

Van Den Eede, G, Aarts, H, Buhk, H-J, Corthier, G, Flint, HJ, Hammes, W, Jacobsen, B, Midtvedt, T, Van Der Vossen, J, Von Wright, A, Wackernagel, W & Wilcks, A 2004, 'The relevance of gene transfer to the safety of food and feed derived from genetically modified (GM) plants, *Food and Chemical Toxicology*, vol.42, no.7, pp 1127-1156, viewed 5th September 2010, Scopus.

Wu, F 2004, 'Explaining public resistance to genetically modified corn: An analysis of the distribution of benefits and risks', *Risk Analysis*, vol.24, no.3, pp 715-726, viewed 8th September 2010, Wiley InterScience.

Ye,X, Al-Babili, S, Kloeti, A, Zhang, J, Lucca, P, Beyer, P & Potrykus, I 2008, 'Engineering the provitamin A (beta-Carotene) biosynthetic pathway into (carotenoid-free) rice endosperm, *Science*, vol. 287, no.5451, pp 303-305, viewed 18th September 2010, JSTOR.

Zhou, Y & Kastenberg, WE 2006, 'Quantification of changes in chemical pesticide human health risk following the introduction of Bt cotton and herbicide-tolerant soybean: A case study', *Human and Ecological Risk Assessment*, vol.12, no.5, pp 871-887, viewed 7th September 2010, ProQuest.

Appendix A - Developing a genetically modified organism

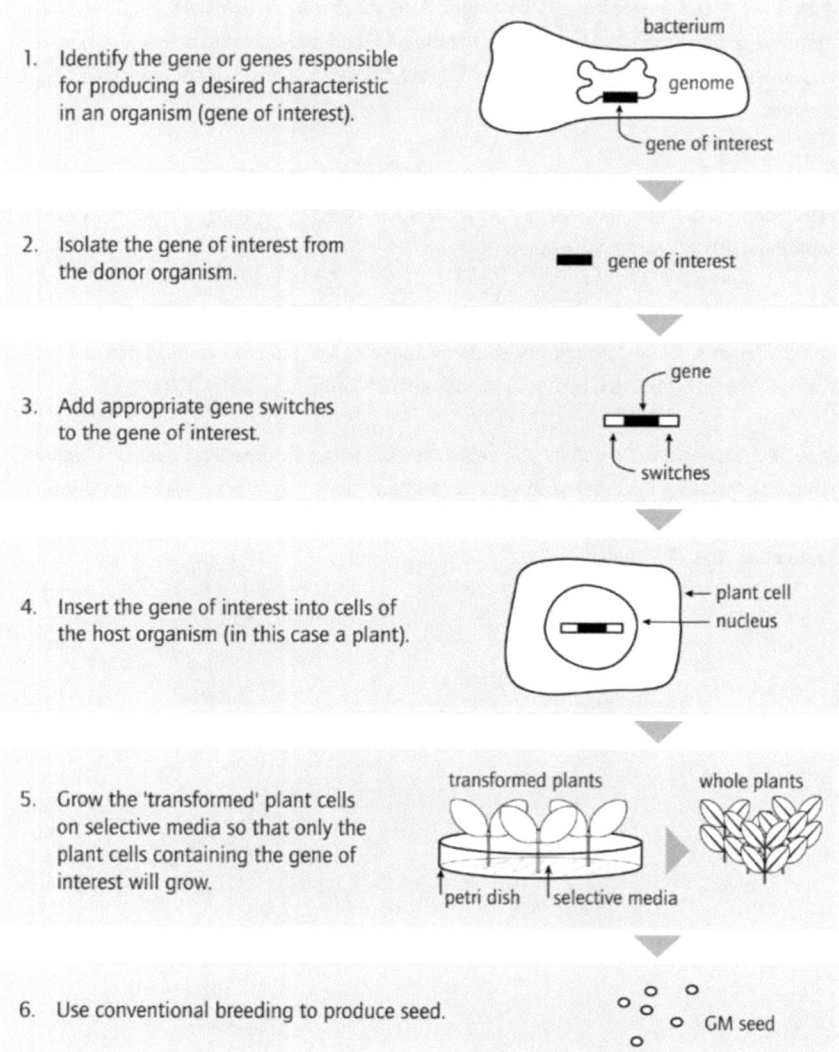

1. Identify the gene or genes responsible for producing a desired characteristic in an organism (gene of interest).

bacterium
genome
gene of interest

2. Isolate the gene of interest from the donor organism.

gene of interest

3. Add appropriate gene switches to the gene of interest.

gene
switches

4. Insert the gene of interest into cells of the host organism (in this case a plant).

plant cell
nucleus

5. Grow the 'transformed' plant cells on selective media so that only the plant cells containing the gene of interest will grow.

transformed plants
whole plants
petri dish   selective media

6. Use conventional breeding to produce seed.

GM seed

Source: FSANZ 2005